Enhancing undergraduate chemistry laboratories

Pre-laboratory and post-laboratory exercises

John Carnduff and Norman Reid

RS·C

Enhancing undergraduate chemistry laboratories
– pre-laboratory and post-laboratory exercises

Written by John Carnduff and Norman Reid

Edited by John Johnston

Designed by Imogen Bertin

Published and distributed by Royal Society of Chemistry

Printed by Royal Society of Chemistry

Copyright © Royal Society of Chemistry 2003

Registered charity No. 207980

For further information on other educational activities undertaken by the Royal Society of Chemistry write to:

Education Department
Royal Society of Chemistry
Burlington house
Piccadilly
London W1J 0BA

Information on other Royal Society of Chemistry activities can be found on its websites:
http://www.rsc.org
http://www.chemsoc.org
http://www.chemsoc.org/LearnNet contains resources for teachers and students from around the world.

ISBN 0–85404–378–0

British Library Cataloguing in Publication Data.

A catalogue for this book is available from the British Library.

Front cover illustration supplied by Corbis.

Foreword

Laboratory work is an integral part of higher education chemistry courses. Those with responsibility for organising such work seek to maximise the students' gain from the time spent in the laboratory. Some class organisers have developed learning tasks to be completed before and after laboratory sessions to make laboratory experiences more meaningful.

In this booklet we attempt to survey materials that already exist and to show the range of strategies and expectations. We have included a set of examples of paper-based material and offered some guidance in constructing such pre-laboratory and post-laboratory exercises. We hope that these may be useful and will encourage those who wish to design their own material. The overall aim is to enrich students' experiences in laboratories.

RS•C

Contents

Acknowledgements

This report is largely based on the work, ideas and experience of many other chemists.

We are grateful to all those who responded to the questionnaire, to letters and in workshops and all those who sent pre-laboratory examples or other useful materials, including:

Dr Dave Adams (Heriot-Watt University, Edinburgh)
Dr Edgar Anderson (University College London)
Dr Stephen Breuer (Lancaster University)
Prof Lois Browne (Alberta, Canada)
Dr Bill Byers (University of Ulster)
Dr Bob Bucat (Western Australia)
Dr Odile Gallais (Paris Sud, Orsay)
Dr John Garratt (University of York)
Dr Caroline Koh (King's College London)
Prof Ray Matthews (University of North London)
Dr David McGarvey (Keele University)
Dr George McKelvy (Georgia Tech, Atlanta, USA)
Prof Liam Murphy (University College Cork, Ireland)
Dr Tina Overton (University of Hull)
Dr Peter Preston (Heriot-Watt University, Edinburgh)
Dr Alan Scotney (University of Glasgow)
Dr Tim Wallace (University of Salford)
Prof Paul Walton (University of York)
Dr Hazel Wilkins (Robert Gordon University, Aberdeen)
Dr Janice Aldrich-Wright (Western Sydney, Australia)
Dr Paul Yates (Keele University)

All colleagues in the chemistry department and in the Centre for Science Education at the University of Glasgow.

All those who commented on the draft collection of pre-laboratory exercises including

Prof Pat Bailey, Dr Stuart Bennett, Prof Alex Johnstone and Dr Tina Overton

Dr Tony Ashmore of the RSC

The Marjorie Cutter Bequest

RS•C

Introduction

Laboratory work is regarded as an essential and integral part of undergraduate chemistry courses and the laboratory provides a setting for training not only in practical hand and instrument skills but also for many of the thinking, planning, recording, interpreting and group working skills that a degree course must include.

However, students often learn very little from the time they spend in the laboratory. They may see or make few connections with lectures. They may feel they are being treated less as adults in first year at university, with its closed circumscribed laboratory work, than they were at school. They may feel that the assessment credit does not match their effort or that the assessment procedure encourages them – though not deliberately – to cut corners or copy. They may feel overwhelmed by the amount of information provided and be unable to see the wood for the trees. More than half of them will have no intention of becoming practising chemists anyway.

Teaching staff are concerned that students often come ill-prepared. There are pressures over costs of man-power, materials, disposal, safety and time; and yet we want – and are being expected – to provide training, skills development and intellectual stimulation.

It is obvious that anything that maximises what students gain from the time they are actually in the laboratory is worth doing. This might include changes in the strategy **during** the laboratory. Increasingly frequently it will involve an expectation or a requirement that the students prepare in their own time **before** the actual experiment.

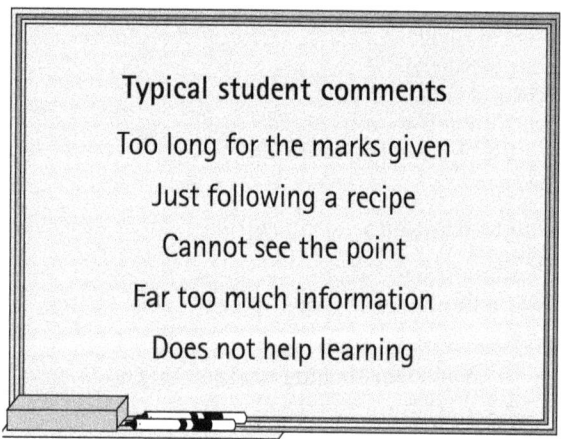

Students are usually expected to read the manual giving detailed instructions and often giving background information on safety, formulae and equations. They may be expected to view videos or CD images of the equipment and how to handle it, to work through computer simulations of the procedure, or of the subsequent data handling, or to practice calculations. Often there is an introductory talk or demonstration. All this helps.

An additional, relatively simple, procedure is to set a pre-laboratory exercise to be submitted and checked before work begins. The exercise can have many aims. Overall, it can stimulate students to think through the laboratory work, with a mind prepared for what will happen. It may require students to recall or find facts such as structures, reaction equations, formulae, definitions, terminology, symbolisms, physical properties, safety hazards or disposal procedures. But it can do much more.

Pre-laboratory work may check that experimental procedures have been read and understood and it can offer practice in data handling, drawings or calculations of the kind to be used in the write-up.

Pre-laboratory work may lead students into thinking about the procedure or the concepts and may encourage students to connect and revise prior knowledge, thus providing some reassurance about their grasp of the topic.

It can involve students in planning (the apparatus, the procedure, the quantities, the data presentation). It can bridge the (common) gap between laboratory and lecture, experiment and application. It can ask for proposals or even offer a challenge. All this is likely to improve motivation and learning.

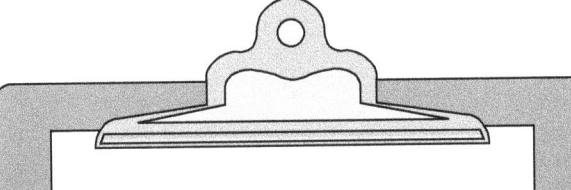

Why use a pre-laboratory exercise?

Private pre-laboratory work by the student might...

- Ensure that background information is recalled
- Connect and revise prior knowledge
- Provide some reassurance to the student about their grasp of the topic
- Check that any procedures have been read and understood
- Practice appropriate data handling, drawings or calculations
- Lead the student into thinking about the procedure or concepts
- Involve the student in planning
- Connect the experiment with other parts of the course
- Relate the experiment to the outside world
- Improve motivation and, perhaps, invite a prediction or offer a challenge.

Such pre-laboratory exercises are being used in a number of chemistry courses in the UK, in North America and a few elsewhere. There is a great variety in length, coverage and management procedures, presumably reflecting a great variety of aims, timing and assessment methods. They have mainly been constructed for introductory or level 1 courses but the strategy could be useful at all levels. Indeed, in a more sophisticated form, such activities would be the normal first stage of an Honours project.

Overall, pre-laboratory exercises are a fairly simple way of preparing the minds of students. They are not expensive to construct and they do help. However, writing them, of course, requires some care. Before attempting the task it is important to specify the aims of the particular experiment. The writer needs to have a good idea about the prior knowledge most students are likely to have (facts, representations, concepts, terminology *etc*) and appropriate links to this knowledge. Experience will also indicate the likely hang-ups and misconceptions that the pre-laboratory exercise could address.

Two other decisions need to be taken. First, the length of the exercise is important; long enough to be worthwhile but not so long that student resistance is generated. Secondly, the procedure for assessment or checking needs to be thought through carefully.

At this stage, the pre-laboratory exercise, the instructions and the post-laboratory report or questions need to be constructed as an a integrated package and designed to fit with the assessment procedure, the other activities and, importantly, the course aims. The process of building on to existing knowledge recognising its limits, acquiring new data, and reflecting and discussing is, of course, always good educational practice.

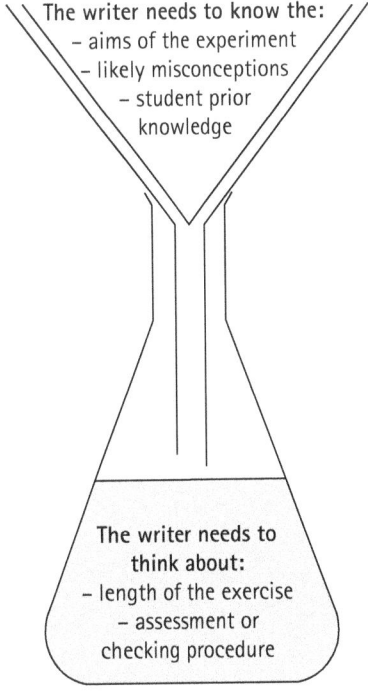

The writer needs to know the:
- aims of the experiment
- likely misconceptions
- student prior knowledge

The writer needs to think about:
- length of the exercise
- assessment or checking procedure

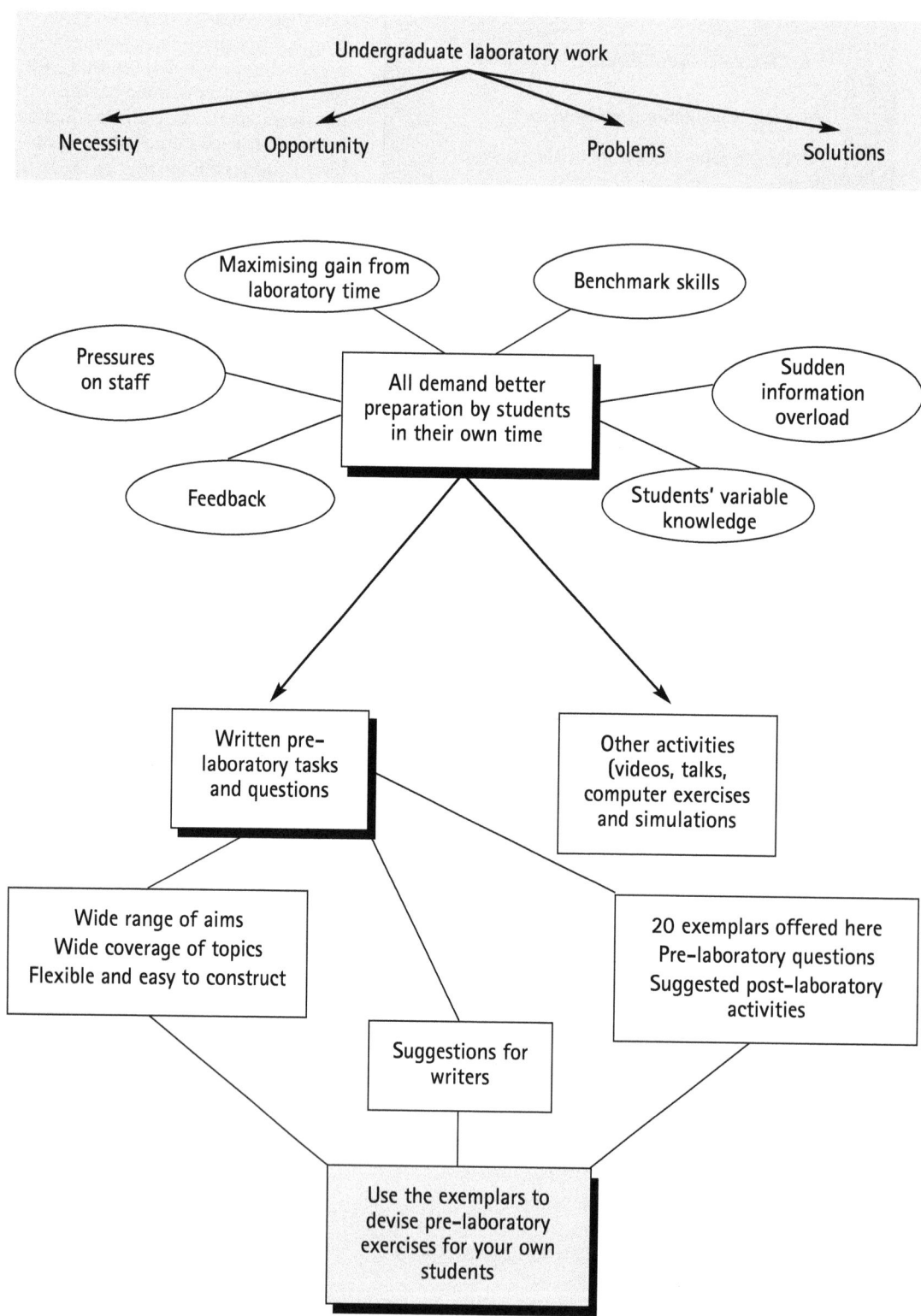

How to use this booklet

The 20 examples of pre-laboratory exercises that follow have been chosen to include typical experiments, rather loosely classified as introductory, level 1 or level 2 and including general, physical, inorganic and organic examples. Some are existing exercises from various universities, kindly and freely offered. Some have been written specifically for this project to go with existing experiments. They have all been edited into a standard layout, but with no implication that this is any better than alternatives.

They have been constructed as paper-based work to be written at home by students who already have read the manual and any other set reading. They can then be checked or marked by a demonstrator before laboratory work starts. Pre-laboratory exercises designed for electronic marking obviously require a different style.

Each pre-laboratory exemplar probably contains more questions than one would use. This is done deliberately to illustrate the range of possible questions. Some authors might wish to include pre-laboratory questions about safety or literature or quantities, say, in all their examples. These appear in only some of the following examples. The emphasis here is more on grasping concepts and experiment design.

Post-laboratory expectations vary enormously from university to university, from lecturer to lecturer. Some expect immediate report presentation for marking while others expect laboratory notes to be written up formally later. Here, the suggested post-laboratory tasks or questions deal largely with implications, applications and extensions, and sometimes connections with other areas of chemistry. Others might prefer to emphasise aspects like data presentation, precision or critical review.

Pre-laboratory exercises can be a part of the range of methodologies that instructors might use to maximise the value gained from student time in the laboratory.The thinking that should go into writing exercises is largely the same as that required in designing other teaching materials. Their aim is to prepare and involve minds, to make connections, to minimise confusion and overload, to raise interest, and to improve achievement.

The examples are just that – examples – to show the range of what might be done. Laboratory organisers may wish to write their own, for **their** experiments, **their** students, **their** programme and timetable, **their** assessment procedures, and **their** aims and priorities. After the examples, there is a review of the use of pre-laboratory exercises, expanding on the topics in this section, followed by references.

Readers will no doubt find among the examples things they would do differently. They will feel they could do better for their laboratories.

This is an invitation to do just that!

> ### Layout of pre-laboratory examples
>
> Experiment – very brief outline
>
> Level
>
> Prior information provided
>
> Aims – briefly stated
>
> Pre-laboratory exercise
>
> Post-laboratory questions
>
> Occasional comments.

> ### Reminder
>
> The exemplars are illustrations only
>
> The aims are not given in detail
>
> The experimental outlines are very brief
>
> The level suggested is only a rough guide
>
> There are numerous questions – selection may be needed
>
> They are not designed to be used without selection and modification

RS•C

This page has been intentionally left blank.

1 The recrystallisation of phthalic acid from water

Level	Introductory	**Aims**	
Prior information	Detailed instructions, databooks, COSHH data	To introduce equipment and encourage thought about procedure and safety. To use published data.	

Pre-laboratory work/questions

1. Set out in flow diagrams the sequence of steps in each of the three slightly different procedures.
2. Prepare the outline of the table you will use to record and compare the outcomes of the three trials.
3. How do you intend to measure the additional volume of water needed to prevent premature crystallisation?
4. After hot filtration, you are asked to wash the filter paper with water boiled in the original flask. Why use this flask?
5. Why should deionised water be used for chemistry experiments?
6. From the solubility data given, calculate the weight of phthalic acid that would remain dissolved in 10 cm^3 of water at 14 °C. Show the calculation.
7. State the hazards and waste disposal method for phthalic acid.

Post-laboratory work/questions

1. Make a record of your results and your conclusions about the best procedure.
2. If a substance is too insoluble in water and too soluble in ethanol to use either as solvent for recrystallisation. What might you try?

Comments
Simple and basic, but strong on planning of procedure and results, deduction from observations.

Source
University of North London, with minor modifications.

2   *The distillation of ethyl acetate*

Level	Introductory	**Aims**
Prior information	Handout sheets with background and outline instructions. Video showing set-up. Reference texts and COSHH data.	To introduce equipment and encourage thought about procedure and safety. To use graphs and consult literature data.

Pre-laboratory work/questions

1. Draw a diagram of the distillation apparatus you will use.
2. Sketch in outline the table you will use to record your observations with headings and units.
3. Sketch a graph to show how you expect the temperature of the condensing vapour (*y*-axis) to vary with time (*x*-axis) during the distillation. Using arrows, label the graph at the points at which you expect to change receiver flasks.
4. Give the literature value for the bpt of ethyl acetate and quote fully the reference.
5. What is the meaning of the terms: a volatile liquid, two miscible liquids?
6. Close-down procedure includes: allowing the flask to cool, switching off the mantle, turning off the water, lowering the mantle, changing the receiver.

   Place these operations in the correct sequence.
7. Why must the following safety instructions be observed?

   Remove the mantle before pouring the sample into the flask.

   Do not leave the distillation unattended.

   Stop when 5 cm^3 remain in the distillation flask.

Post-laboratory work/questions

1. Record your results and conclusions.

Comments

Simple and basic, but strong on planning, safety and recording. Checks terminology and referencing, practices graphs.

Source

University of North London, with minor modifications.

3 Using a separating funnel to separate benzoic acid and camphor

Level	Introductory	Aims
Prior information	Manual with concise **numbered, itemised** instructions. Background information on the compounds and hazards and detailed appendices on techniques and equipment.	To use a separating funnel for acid/base extraction and think about how, why and when it works.

Pre-laboratory work/questions

1. Read the appendices and instructions.
2. What compounds do you expect to have isolated at the end of step 7 and at the end of step 11?
3. How will you decide how much magnesium sulfate to use for drying the solution in steps 6 and 10?
4. To what pH (roughly) should you acidify the aqueous phase in step 8?
5. How will you know whether you have added enough concentrated hydrochloric acid at step 8?
6. Benzoic acid is almost insoluble in water whereas ethanoic acid is soluble. Suggest why.
7. Why is the anion of benzoic acid soluble in water?
8. After the acidification in step 8 you will extract the product with ethyl acetate. How else might the product be isolated?

Post-laboratory work/questions

1. Record the weights and melting points of your products.
2. Write an equation to show how the product isolated after step 11 reacts with sodium hydroxide.
3. Write an equation to show how the conjugate base of this compound reacts with hydrochloric acid.
4. If your camphor/benzoic acid mixture had also contained *para*-toluidine (a basic water-insoluble liquid amine), how would you modify the experiment so as to isolate all three compounds in pure form? What pHs would you need in the aqueous layers at each step?

Comments
The background is all on one page. The short, numbered instructions are all on one page. The details of apparatus and handling (balance, separating funnel, drying etc.) are in the appendices. The pre-laboratory exercise checks that the procedure has been read and understood. Strong on thinking about procedure and alternatives/variants.

Source
Heriot-Watt University, Edinburgh, with minor modifications.

RS•C

4

Preparing and standardising a ~0.1 mol dm⁻³ sulfuric acid solution. Titrating a solution of commercial SinkDblock to estimate its sodium hydroxide content

Level	Introductory	**Aims**
Prior information	School knowledge, handout on safety, precision, indicators and glassware, detailed instructions.	To introduce titration procedure and equipment, moles and concentrations. To consider reproducibility and precision. To connect the laboratory work with the 'real' world.

Pre-laboratory work/questions

1. Write the formulae for sodium hydroxide and sulfuric acid and calculate their molecular weights.

2. Write equations for what happens when each is dissolved in water, and for the reaction when the solutions are mixed. At the equivalence point what is the relationship between the amounts of H^+ and HO^- put into the flask?

3. *SinkDblock* claims to contain '20% sodium hydroxide by weight'. Calculate from this the amount of sodium hydroxide in 1 kg of *SinkDblock* (in grams and in moles)

   You will use a 0.1 mol dm⁻³ solution of sulfuric acid. What concentration of sodium hydroxide (derived from *SinkDblock*) would be appropriate to use for the titration? You will prepare 250 cm⁻³ of such a solution. What weight of *SinkDblock* should you dissolve?

4. What are the hazards of:

   a) skin contact with sodium hydroxide; and

   b) inhaling sulfuric acid droplets?

5. What are the relative precisions of: a pipette, a standard flask, a burette?

Work/questions during or after the laboratory

1. In the standardisation, when do you decide you have reliable results?

2. Calculate the concentration of your acid and state the precision.

3. From your titration results, calculate the sodium hydroxide content of *SinkDblock* as a weight percentage.

4. Compare your result with those of the rest of the class and with the label.

5. Write a brief report of what was done and your conclusions.

6. By what kind of reaction would *SinkDblock* break up fat in the drain?

7. The titration involved a rapid and complete acid/base reaction. What other (different) types of reaction might be used for titrations?

Comments
The pre-laboratory exercise checks knowledge of formulae, molecular weights, moles, equations, hazards. Practises calculations of concentrations. Invites thought on precision, experiment planning. Informally written. Everyday relevance. The post-laboratory questions require

a written account, calculations, group discussion on precision and accuracy and extension to related topics.

Source
Université de Paris Sud, Orsay, with minor modifications.

5

The extraction of spinach pigments, separation by chromatography and estimation by spectrometry

Level	*Introductory*	*Aims*	
Prior information	*Large manual/handout on pigments, quantities, spectra, chromatography, dipole moments, detailed instructions and a spinach tart recipe.*	*To introduce organic compounds, visible spectrometry, polarity and chromatography. To practice safety and handling techniques, units, amounts and concentrations.*	

Pre-laboratory work/questions

1. What is the hazard of inhaling methanol?
2. What is the systematic name for chloroform?
3. What types of substances are present in 40–60 petroleum? Can it damage skin?
4. What is the industrial source of chlorophyll?
5. What is meant by the phrase 'vitamin A precursor'? What weight of spinach in your diet would provide the daily requirement for vitamin A?
6. Put the solvents listed in order of increasing dipole moment: water, methanol, 40–60 petroleum, chloroform.
7. Decide, from their structures, which spinach pigments are polar? Which will dissolve readily in aqueous methanol?
8. In what wavelength ranges do carotene, chlorophyll and xanthophyll each absorb strongly?
9. From the spectra provided, determine the λ_{max} and ε for carotene.
10. Draw a flow diagram for the experiment procedure.

Work/questions during and after the laboratory

1. Decide which layer is which in the separating funnel.
2. Decide, on the basis of colour and dipole moment, which pigment is in which layer.
3. What happens to the anhydrous sodium sulfate you used and why?
4. Which part of the chlorophyll molecule can be hydrolysed and why does this change the solubility?
5. Determine ε from the spectra provided and so deduce the number of moles of pigment in the weight of spinach you used and so deduce the concentration in the spinach in mol kg^{-1} and mg kg^{-1}. This value is likely to be much lower than the actual concentration. Suggest why.

Comments
Checks knowledge, understanding and interpretation of the handout, hazards, nomenclature and costs; experiment planning; decision taking; everyday relevance; short direct questions throughout the experiment; calculations. Requires evaluation of the result. No formal report.

Source
Université de Paris Sud, Orsay, with minor modifications.

RS•C

6

The preparation of crystals of the complex $K_3[Cr(C_2O_4)_3].3H_2O$ from oxalic acid and potassium dichromate

Level	One	Aims
Prior knowledge	Lectures and a previous redox titration experiment. Instructions.	To explore transition metal complexes and use ion-electron equations for quantitative calculations on redox reactions.

Pre-laboratory work/questions

1. Use a textbook or dictionary to give definitions of the following chemical terms. Explain how the meaning connects with the word in brackets
   a ligand(ligature), a hydrate(dehydrated), transition metal(transfer), bidentate (dentist) and octahedral(octave)

2. Give an example of a naturally occurring complex of a transition metal with a polydentate ligand.

3. Draw the structure of the (planar) oxalate ion. How many planes of symmetry divide it?

4. The Cr^{3+} ion combines with three bidentate oxalate ions to form an octahedral complex anion. Draw this ion and work out its charge.

5. What is the O–Cr–O angle and the O–C–C angle in the rings of the complex?

6. Write out the ion-electron half equations for the oxidation of oxalate ion to carbon dioxide and the reduction of $Cr_2O_7^{2-}$ to Cr^{3+} in the presence of H^+.

7. Work out the equation for the redox reaction of oxalate ion with dichromate ion in acid solution.

8. What is the oxidation state of chromium in the ion $Cr_2O_7^{2-}$?

9. Work out the theoretical weight of $K_2Cr_2O_7$ needed to prepare 5.0 g of $K_3[Cr(C_2O_4)_3].3H_2O$. You will use this weight for the preparation.

10. In step 4 of the procedure you will add ethanol to the aqueous reaction mixture. What effect will this have on the solubility of the ions?

Post-laboratory work/questions

1. Use your pre-laboratory calculation to work out your yield. Suggest two possible reasons why it is not 100%.

2. Suggest why, when you added the ethanol, the solution became deep purple?

3. Look at the model of the complex ion. How many planes of symmetry divide it?

4. What conclusion can you draw from the experiment about the reaction
   $Cr^{3+} + e^- \rightarrow Cr^{2+}$?

5. The ratio of complex ion to potassium ion in the crystals is fixed (1:3) by the charges. The ratio of ions to water is also fixed. By what?

Comments
Covers terminology, real world, moles and equations. Connects with lectures, other experiments, 3D concepts and symmetry, crystal packing. Draws conclusions from observations. Followed by analysis of the crystals by titration of liberated oxalate ion with permanganate.

Source
University of Glasgow, with minor modifications.

7
Discovering pHs of acids and salts using Universal indicator. Following titrations of weak and strong acids with a pH meter

Level	*Introductory*	*Aims*	
Prior information	*Nothing except manual and textbook.*	*To explore simple ideas about acids and bases, extent of dissociation and pH titration curves and to practice mole calculations.*	

Pre-laboratory work/questions

1. Write the formulae for nitric acid, ammonium nitrate, sodium hydrogencarbonate, ethanoic acid and sodium ethanoate. Draw out fully the structures for hydrogencarbonate and ethanoate ions.

2. Write equations for the reactions in which:
   a) nitric acid dissociates to give H^+
   b) ammonium ion dissociates to give H^+
   c) hydrogencarbonate ion dissociates to give H^+
   d) hydrogencarbonate ion combines with H^+
   e) ethanoic acid reacts with sodium hydroxide.

3. Use the definition of pH and your calculator to work out:
   a) the $[H^+]$ in a solution of pH 3.00
   b) the $[H^+]$ in a solution of pH 3.30
   c) the pH of a solution of $[H^+] = 2 \times 10^{-9}$ mol dm^{-3}
   d) the pH of a solution of $[H^+] = 4 \times 10^{-9}$ mol dm^{-3}

4. If 10 cm^3 of a solution of HCl reacts exactly with 9.8 cm^3 of a 0.12 mol dm^{-3} solution of NaOH, work out the number of moles of HCl in the 10 cm^3 sample. Now find the concentration of the HCl.

5. The pH of a 0.10 mol dm^{-3} solution of HCl is found to be 1.0 whereas the pH of a solution of another acid of the same concentration was found to be 3.0. What does this tell you about the extent to which each acid is dissociated?

6. When testing the pH of unknown solutions, why is it wise to add indicator to a little water in a test-tube and then add the unknown solution, rather than the other way round?

Post-laboratory work/questions

1. Make a table of the pH you found for each of the test solutions and write equations to account for the results.

2. Label each of the test substances in your table as strong acid, weak acid... *etc.*

3. Using your results, predict the pH you expect to find for solutions of each of the following – sulfuric acid, ammonium nitrate, potassium sulfate and sodium carbonate.

4. What substance will be present in solution at the exact equivalence point in your titration? Use your data to predict the pH at the equivalence point in your titration.

5. Plot a graph of pH against volume of 1.00 mol dm^{-3} NaOH added. Mark the equivalence point and the required volume on your graph, and so work out the exact concentration of your acid. Compare the graphs for the two acids.

6. Why are the pHs at the beginning of the two titrations different?

7. Why are the pHs at the equivalence point of the two titrations different?

8. Why are the two graphs virtually identical after the equivalence point?

Comments
The pre-laboratory exercise checks knowledge of formulae and equations, using a calculator, appreciation of log scale of pH, simple mole calculations, deduction of extent of dissociation from observation and one aspect of experiment planning. Post-laboratory exercises require collaboration and discussion between students, using

observed data to make predictions on analogous cases, thought about extent of H^+ ion transfer, using observed data to determine equivalence points and extending pre-laboratory mole calculations to determine concentration.

Source University of Glasgow, with minor modifications.

RS•C

8 pH, Buffers and the Henderson equation (Tests and both titrations as in example 7)

Level	*One*	*Aims*
Prior information	*Lectures on H^+ transfer equilibria, manual and textbook.*	*To understand some applications of the Henderson equation for HA/A^- mixtures and to use a pH meter.*

Pre-laboratory work/questions

1. Write out the reaction equation for dissociation of a weak acid HA .
2. Write out the definition of K_a for this acid.
3. Write the Henderson equation of the pH of a mixture of HA and A^-.
4. If a little HO^- is added to a solution of HA, what will happen to;
      a) [HA]   **b)** $[A^-]$ **c)** K_a **d)** $[H^+]$
5. If a little A^- is added to a solution of HA, what will happen to:
      a) [HA]   **b)** $[A^-]$ **c)** K_a **d)** $[H^+]$
6. The pH of a dilute solution of NaA is found to be 8.2. Deduce $[H^+]$ and use the chemical equilibrium equation to explain the high pH.
7. If a 0.1 mol dm^{-3} solution of HA has pH 2.90, calculate K_a for this acid and the extent of dissociation at this concentration. If the solution is diluted what will happen to the extent of dissociation?
8. Outline the advantages and disadvantages of
      a) a single indicator   **b)** Universal indicator   **c)** pH meter
   for determining the pH of a solution of vinegar or beer
   and for determining the equivalence point in the titration of HA with NaOH.

Post-laboratory work/questions

1. Draw graphs of pH against volume of NaOH added. Using your observations on the pH of salt solutions, estimate the pH of the equivalence point in each titration.
2. In the HCl titration, the pH rises from about 1 to about 2 over the first 90% of the titration. Explain why.
3. For weak acids, the Henderson equation implies that, at the stage when half the acid has been converted to salt and half remains, pH = pK_a. From your graph deduce pK_a for ethanoic acid. Use this value to predict the pH at 20% and 80% conversion and compare the results with your observations.
4. How would the pH change if, at the 20% stage, the mixture was diluted with water to twice its volume?
5. What does the Henderson equation tell you about the gradient of the graph between 20% and 80% conversion?
6. Suggest a way of preparing a buffer solution of pH 5.2 starting with sodium acetate and hydrochloric acid.
7. What would you use to prepare an effective buffer solution of pH 8.8?
8. If a titration were monitored using a single indicator, HIn, (yellow) which is converted to an anion, In^-(blue), and pK_a of HIn is 6.0, what colour would the solution be at pH 3, 5, 6, 7?
9. Would this be a suitable indicator for titration of
      a) ethanoic acid with NaOH   **b)** hydrochoric acid with NaOH?

Comments	*Source*
The pre-laboratory revises basic ideas about solutions of HA, A^- or both. It also requires thought about methods. Post-laboratory work requires deduction, interpretation and extension from observations and some lateral thinking.	*University of Glasgow*

9 The dehydration of tert-amyl alcohol catalysed by concentrated sulfuric acid and the collection of 2-methyl-2-butene by slow distillation below 40 °C, and tests on its reactions

Level	One	Aims
Prior information	Lectures, manual with structures, safety information and procedure.	To apply knowledge of naming, structures and mechanisms, and to perform a simple distillation and understand the procedure.

Pre-laboratory questions

1. Give the systematic name of *tert*-amyl alcohol.
2. Write a reaction mechanism for the formation of 2-methyl-2-butene from *tert*-amyl alcohol.
3. Does your mechanism fit with the suggestion that the acid is a true catalyst?
4. What other alkene might be formed in this process?
5. Why is 2-methyl-2-butene the favoured product?
6. Draw the structure of a third branched alkene, C_5H_{10}.
7. What are the molecular formulae and molecular weights (to 1 decimal place) of 2-methyl-2-butene and *tert*-amyl alcohol?
8. Given that the density of *tert*-amyl alcohol is 0.805 g cm^{-3}, calculate the mass and number of millimoles in the 10 cm^3 of alcohol you will use.
9. Calculate the mass of alkene expected if you were to obtain a 100% yield.

Post-laboratory work/questions

1. Suggest reasons why your yield was less than 100%.
2. Dehydration of alcohols to alkenes is potentially reversible. Suggest why, never the less, you got a good yield of alkene.
3. Reaction of the same alcohol with concentrated (10 mol dm^{-3}) hydrochloric acid converts it to *tert*-amyl chloride. Suggest a mechanism for this conversion and a reason why the alkene is not formed.
4. You carried out the addition of bromine to your alkene. What would be the product of addition of bromine to cyclohexene? Write out the mechanism.
5. What modifications of the procedure would be appropriate for dehydration of
   a) propan-2-ol
   b) 2-phenylpropan-2-ol?

Comments
Requires a yield calculation and thought on procedure, mechanisms, product selection, isomers and nomenclature.

Source
Heriot-Watt University, Edinburgh

RS•C

10
The alkaline hydrolysis of the ethyl ester of an unknown solid acid. Isolation of the acid and identification by mp comparison against a list of substituted benzoic acids provided. Different students have different esters.

Level	*One*	*Aims*
Prior information	*Lectures on physical and organic topics and a manual with detailed, **numbered** instructions.*	*To carry out the hydrolysis, isolate and identify the acid* *To think about physical properties, mechanisms and factors affecting rate and to make comparisons among students.*

Pre-laboratory work/questions

1. Draw in full the structures of ethanoic acid, benzoic acid, ethyl ethanoate and ethyl benzoate. Explain why, of these four, only ethanoic acid is soluble in water.

2. Write out the mechanisms of all four steps in the conversion of the ester to the acid. Which step is the slowest one? Which corresponds to stage 4 of the procedure?

3. Look up the pK_a values for HCl, HOC_2H_5 and HOCOR and so explain why the proton transfers you have drawn are likely to be complete.

4. Assuming your ester is insoluble in hot water, how will you tell when hydrolysis is complete?

Post-laboratory work/questions

1. Record the mp, literature mp, name and structure of your acid.

2. Why are all the listed benzoic acids odourless solids, whereas many of the esters are liquids?

3. Other derivatives of acids can be hydrolysed by the same mechanism. Why are esters hydrolysed faster than amides but slower than anhydrides?

4. Why is the ester hydrolysis faster at 100 °C than at 20 °C, faster at pH 14 than pH 7 and faster when bubbling vigorously?

5. Where did the ethanol formed end up? How would you modify the procedure to hydrolyse octyl benzoate and isolate both products?

6. Did any of the esters dissolve readily at the beginning? If so, why? (Some students had salicylates.)

Comments	*Source*
The exercise builds on several areas of previous lectures, requires deductions from literature data and observation, thinking about procedure and opportunities for group comparisons.	*Heriot-Watt University, University of Edinburgh and University of Glasgow*

11

(a) The microscale preparation of benzyl ethanoate by reaction of benzyl alcohol with excess ethanoyl chloride in petrol
(b) The preparation of ethyl benzoate similarly by reaction of benzoyl chloride with excess ethanol
Interpretation of spectra

Level	One	*Aims*
Prior information	*Lectures on esterification and on infrared and ^1H NMR spectra. Video on bp determination and textbook references. Instructions.*	*To practice using spectra. To use acid chlorides safely. To think about reactivity and procedure.*

Pre-laboratory work/questions

1. Draw the structures of the two esters.
2. Outline two other methods of converting alcohols to their ethanoates.
3. Why is the carbonyl group in the products likely to cause strong absorption in the infrared?
4. At roughly what wavenumber do you expect this absorption for the two products?
5. Look up the ^1H chemical shifts typical of the methyl groups of – propane, dimethyl ether, propanone and toluene (methylbenzene), and briefly account for the differences.
6. Predict the ^1H spectra of the two products in terms of chemical shift, integral, multiplicity and coupling constant.

Postlaboratory work/questions

1. What was the gas evolved in the reactions?
2. Suggest why the procedures for preparing the two esters are significantly different.
3. How might you check whether your crude ester contains unreacted alcohol?
4. Contrast your infrared spectra, and the NMR spectra provided, with one another and with your predictions.
5. Compare your observed boiling points with literature values.
6. How does the procedure for making benzyl ethanoate ensure the product is free of acid chloride and acid?
7. What does the workup procedure tell you about the rate of hydrolysis of the esters?

Comments
Includes revision and predictions, interpretation of the procedure and some lateral thinking.

Source
Lancaster University, with minor modifications.

12

The temperature dependence of the chemiluminescent reaction in lightsticks between aryl oxalates, alkaline hydrogen peroxide and a fluorescent dye, involving ICT skills

Level	One	Aims
Prior information	A copy of a textbook account (B.Z. Shakashiri, Chemical demonstrations. Wisconsin: University of Wisconsin Press) of this topic, including reaction equations, some background information and references. ICT training and lectures on organic and physical topics.	To investigate and interpret how temperature affects the intensity of the light emitted. To practise library and ICT skills. (Students are given a detailed list of objectives and of skills to be enhanced.)

Pre-laboratory work/questions

1. Prepare an experimental plan involving the photodiode light meter and the water bath, to investigate the influence of temperature on the luminescence intensity of the lightstick reaction and so obtain the 'apparent' activation energy (see P.W. Atkins, *Physical chemistry*. Oxford: OUP.). Carry out a COSHH risk assessment.

2. Plan how to handle your data on an *Excel* spreadsheet and to construct an Arrhenius plot and so deduce the apparent activation energy.

3. Using *ISIS Draw* or *ChemDraw*, draw plausible mechanisms, with curly arrows, for the reactions in the textbook account that lead to the cyclic intermediate.

4. From the textbook account, quote:
   a) the names of the authors of the symmetry rules relating to the decomposition of the cyclic intermediate;
   b) the experimental evidence supporting the electron transfer luminescence mechanism;
   c) the wavelength of maximum fluorescence intensity from 9,10-bis(phenylethynyl)anthracene; and
   d) the colour of light this corresponds to.

5. Using a suitable search engine (*Google, Yahoo, AltaVista*) search the WWW for a definition of chemiluminescence. Quote it and give the web address.

6. Locate in the library the journal article given as ref 9 in the textbook account. Photocopy the first page and write a two sentence summary of the abstract.

Post-laboratory work/questions

1. Report briefly on your procedure.

2. Input your data on *Excel* and plot *I* versus *T*. Manipulate your data within the spreadsheet (using formula entries) to obtain data forming the basis of an Arrhenius plot. Construct the plot, fit a trend line and, from the equation of the line, deduce the apparent activation energy.

3. Compare the values for lightsticks with different dyes and interpret your conclusions.

Comments

An exercise designed for a small class. Many ICT and communication skills associated with an intriguing phenomenon and a simple procedure. Minimal instruction beyond the textbook account. Students work in pairs and design the experiment and data handling.

Pre-laboratory work checks reading, understanding and ICT competence and makes connections with organic courses.

Source

Keele University, with minor modifications.

13 The preparation, infrared and electronic spectra of the complex bis(2,4-pentanedionato)copper (II) [Cu(pd)₂]

Level	Two	Aims
Prior information	Lectures, textbooks, manual with information and instructions for preparation and spectrometry.	To practice representations of orbital energies and interpret infrared and visible spectra. To record infrared spectra.

Pre-laboratory work/questions

1. Make sure you understand the terms ligand, bidentate and chelate.
2. Draw and label a diagram showing how the d-orbitals split in an octahedral field.
3. Draw a diagram showing how the orbital energies change as an octahedral complex is distorted towards square planar geometry.
4. Write out the electron configuration of the free Cu^{2+} ion.
5. In the diagram you have drawn above, indicate how the d-electrons would be arranged in:
   a) an octahedral complex of Cu^{2+} and
   b) a square planar complex of Cu^{2+}
6. How many absorption bands would you expect in the electronic spectrum of:
   a) an octahedral complex of Cu^{2+} and
   b) a square planar complex of Cu^{2+}

Label your diagrams.

Post-laboratory work/questions

1. Discuss differences between the infrared spectra of pentanedione and its copper complex.
2. Explain the origin of the three bands in the electronic spectrum of the square planar complex.
3. Explain the changes in the electronic spectrum you saw after addition of pyridine.

Comments

The pre-laboratory exercise checks terminology, bookwork knowledge and requires application of concepts and prediction. The post-laboratory questions require interpretation of the observations. Strong on theory, but it might usefully have included some applications

Source

University of Glasgow

14

The infrared spectra and reactions of $[Co(NH_3)_6]Cl_3$ prepared from $[Co(H_2O)_6]^{2+}$ by addition of NH_4Cl, then NH_3, then oxidation

Level	Two	Aim
Prior information	Textbooks and a manual with detailed instructions.	To exemplify redox reactions and ligand transfers of transition metal complexes. To apply theory on vibrational spectroscopy.

Pre-laboratory work/questions

1. Why is the ligand exchange $[Co(H_2O)_6]^{2+} + 6\,NH_3 \rightarrow [Co(NH_3)_6]^{2+} + 6\,H_2O$ essentially complete?

2. Given the redox values shown, what might happen:
   a) if a solution of $[Co(NH_3)_6]^{2+}$ were exposed to air and
   b) if $[Co(H_2O)_6]^{3+}$ were prepared in solution in water?

$$[Co(H_2O)_6]^{3+} + e^- \rightarrow [Co(H_2O)_6]^{2+} \quad E_o = 1.84\,V$$
$$O_2 + 4H^+ + 4e^- \rightarrow 2H_2O \quad E_o = 1.23\,V$$
$$[Co(NH_3)_6]^{3+} + e^- \rightarrow [Co(NH_3)_6]^{2+} \quad E_o = 0.10\,V$$

3. The four distinct vibration modes for the ammonia molecule are shown in the manual together with their wavenumbers. Do you expect to see all four absorption bands?

4. For each one, how do you expect the wavenumber to change if the ammonia is coordinated to Co^{3+}. What additional bands will appear in the spectrum of the complex $[Co(NH_3)_6]^{3+}$? Explain.

Post-laboratory work/questions

1. Explain the colour change you see when NH_4Cl is added to $CoCl_2.6H_2O$.
2. Explain what you see when concentrated HCl is added to $Co(NH_3)_6Cl_3$.
3. Adding silver nitrate solution to a solution of $Co(NH_3)_6Cl_3$ in water gives a precipitate. What does this suggest about the binding of the chlorine in this complex?
4. Assign the infrared bands you found for $Co(NH_3)_6Cl_3$ and compare them with your predictions.
5. When boiled with NaOH solution, the $Co(NH_3)_6Cl_3$ gave a precipitate of $Co(O)OH$. Suggest two reasons why the ammonia becomes separated from the cobalt.

Comments
The pre-laboratory exercise requires understanding, interpretation and prediction. The post-laboratory exercise includes a check on predictions made in the pre-laboratory exercise and interpretation of observations.

Source
University of Glasgow

15   *Using a pH meter to follow the pH as HCl is gradually added to solutions of acetate ion and histidine anion*

Level	*Two*	*Aims*
		To practise mole calculations. To use the Henderson
Prior information	Level one lectures. An experiment	equation to interpret pH titration curves and buffer
	on simple titration curves, K_a and	mixtures.
	the Henderson equation.	

Pre-laboratory work/questions

1. Write out the Henderson equation for the pH of a buffer mixture of a weak acid (HA) and its anion (A⁻). If you add HCl gradually to a solution of A⁻ and graph the pH against volume added, how can you tell **from the graph** when all the A⁻ has been converted to HA?

2. How, **from the graph**, could you determine the pK_a of HA?

3. Write a series of structures showing what changes will occur when the anion of histidine is treated

   gradually with excess HCl. Note that the H_2N- group is the most basic, the $-CO_2^-$ the least basic and the $-N<$ group of the ring not basic at all.

4. What is the name of the ring system present in the side chain of histidine and why is the $-N<$ not basic?

5. The method requires you to prepare **a)** 100 cm³ of 0.15 mol dm⁻³ acetate ion using solid $CH_3CO_2Na.3H_2O$ and **b)** 100 cm³ of a solution made from solid $C_6H_9N_3O_2.HCl. H_2O$ (15 mmoles) and NaOH (45 mmoles). Work out the weight of each you should use.

   Which two anions will then be present in solution **(b)**?
   If 50 cm³ of this solution is treated with 1.0 mol dm⁻³ HCl, what volume will you need to add to protonate the histidine completely? In fact you will add 30 cm³. Roughly what will the pH be then?

Post-laboratory work/questions

1. Graph pH against volume of HCl added for both titrations.

2. Deduce from the first graph the pK_a of acetic acid.

3. Mark on the second graph the points at which you expect only one form of histidine to be present and then deduce the pK_a values of the primary ammonium and the imidazolium groups of histidine.

4. What can you tell from the graph about the pK_a of the CO_2H group. Once you have estimated this value comment on why it is different from the pK_a of acetic acid.

5. For a histidine unit, **in a protein**, in solution in water at pH 7.0, which nitrogens of the histidine do you expect to be protonated and to what extent?

6. Will both enantiomers of histidine have the same pK_a in aqueous solution?

7. If you had added 40 mmole instead of 45 mmole of NaOH, what difference would it have made to the graph or to the pK_a values?

Comments
Revision and connection with other courses.
Interpretation and applications.

Source
University of Glasgow

RS•C

16
Using the voltage of a concentration cell to determine the silver ion concentration in a saturated solution of AgCl in aqueous KCl and so determine the solubility of AgCl in water

Level	*Two*	*Aims*	*To use and understand concentration cells and to apply the concepts of the Nernst equation and solubility product.*
Prior information	*Lectures on these topics, and a manual with instructions and diagrams*		

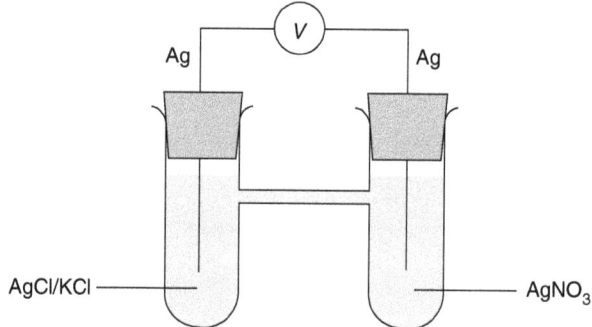

Pre-laboratory work/questions

1. What is the usual test for the presence of traces of chloride ion in a water sample? After any positive test of this kind what is the relationship between $[Ag^+]$ and $[Cl^-]$ in the solution?

2. Guess roughly the solubility of AgCl in water $1g\ dm^{-3}$, $1\ mg\ dm^{-3}$, $1\ \mu g\ dm^{-3}$ or $1\ ng\ dm^{-3}$

3. Write the general form of the Nernst equation for the voltage of an electrochemical cell.

4. For the cell in the diagram, write the ion-electron equations at the left and right electrodes and so decide on the values of E_o and n. Which electrode do you expect to be positive and why?

5. Rewrite the equation for the voltage of this cell in terms of concentrations (you can assume activity coefficients are close to 1 and essentially equal for the two solutions).

6. To find $[Ag^+]_{left}$, what values in your equation will you have to
   a) look up   b) fix   c) measure?

7. To determine K_{sp} for silver chloride, what other value must you know?
   The procedure requires that the left hand tube contains $0.10\ mol\ dm^{-3}$ KCl to which a **tiny** unknown amount of $AgNO_3$ has been added.
   a) Why is this enough?
   b) What complication would arise if you added more?
   c) Why is knowing the amount unimportant?

8. If you double the concentration of silver nitrate in the right-hand tube, what effect will this have on the voltage?

Post-laboratory work/questions

1. Tabulate the values of $[Ag^+]_{left}$ deduced from experiments with various $[Ag^+]_{right}$ values.

2. From the average, determine K_{sp} for silver chloride and estimate the precision of your answer.

3. Use this to work out the solubility of silver chloride in water in $mol\ dm^{-3}$ and in $g\ dm^{-3}$.

4. Why was the high solubility of silver nitrate in water important for this experiment?

5. Suggest why solid silver chloride is so much less soluble in water than solid silver nitrate.

Comments	*Source*
Strong on understanding experiment design.	*University of Glasgow*

17 The redox reaction of periodate with excess iodide at pH 7 and at pH 2, monitored by titration with thiosulfate

Level	Two	Aims
Prior information	Previous experience of iodine/thiosulfate titration and redox concepts, and a manual with instructions, quantities and formulae.	To explore the redox reactions of iodine and to interpret their stoichiometry and pH dependence.

Pre-laboratory work/questions

1. Look up the textbook to find which simple oxoanions, IO_x^-, exist.
2. Work out the oxidation states for the iodine in each of these and give examples of iodine in any other oxidation states you know.
3. Write balanced ion/electron equations for:
   a) the reduction of periodate ion to iodate ion
   b) the reduction of periodate ion to iodine
   c) the oxidation of iodide ion to iodine
4. Predict how the ability of periodate ion to act as an oxidising agent depends on the pH and whether the ability of iodide ion to act as a reducing agent depends on the pH.
5. When solutions of iodate ion and iodide ion in water are mixed, there is no colour change, but when sulfuric acid is then added, the solution becomes brown. Interpret this observation.
6. Write the balanced equation for the reaction of iodine with thiosulfate ions, showing the **structures** of all reagents and products.
7. How many times more hydrogen ions are there per litre at pH 2 compared to pH 7?
8. If you want to make a HA/NaA buffer of pH 7, what should be the pK_a of HA?

Post-laboratory work/questions

1. At pH 7, the iodine formed all comes from oxidation of iodide. From the volume of standard thiosulfate needed and your weight of potassium periodate, work out what redox reaction happened.
2. At pH 2, further iodine was formed. From the further amount of thiosulfate needed, confirm what redox reaction happened at this pH.
3. Why was the borate ion in the buffer not oxidised?
4. Why was the chloride ion, from the HCl that you added, not oxidised?
5. Would perchlorate ion be a more powerful oxidant than periodate?

Comments

The pre-laboratory exercise prepares students for the complexities of the eventual mole calculations. It also requires some revision, some literature searching, some prediction and some recognition of the experiment design.

Source

University of Glasgow

18

To determine the kinetics of decomposition of benzenediazonium ion (PhN_2^+) in water by monitoring the volume of nitrogen evolved against time using a gas burette

Level	Two	Aim
Prior information	Level 1 lectures and a manual with instructions and diagrams of apparatus.	To apply concepts about first order reactions; to consider the experimental strategy and treatment of results; and to consider the implications of the results and the possible extensions.

Pre-laboratory work/questions

1. Write balanced chemical equations for:
   a) the formation of benzenediazonium ion (PhN_2^+) from aniline, nitrite ion and hydrogen ions; and
   b) the hydrolysis of the PhN_2^+ giving phenol and nitrogen gas.
2. Write the universal equation that relates rate of reaction to concentration of reagents A and B.
3. Explain why the hydrolysis reaction, in dilute solution in water, is likely to obey the equation:
   Rate = $k[PhN_2^+]$
4. What are the dimensions of k in this case?
5. For such a first order process, write the integrated rate equation that relates the ratio ($[A]_0/[A]_t$) to the time lapsed (t) and a constant (k).
6. We can take advantage of the fact that we do not need to know the absolute values of $[A]_o$ or $[A]_t$, but only the value of their ratio. Also, for this reaction, it is much easier to measure the amount of nitrogen collected at time t (V_t)   (proportional to amount of ion that has decomposed) and the amount finally collected (V_f)   (likewise proportional to the amount of ion at the start) rather than to measure the concentration of the reactant ion.

   The difference ($V_f - V_t$) will be likewise proportional to the amount of the ion still present.

   Satisfy yourself that the equation $\ln \dfrac{V_f}{V_f - V_t} = kt$ is equivalent to the one you wrote above.

7. Rewrite this equation in the form $y = mx + c$.   (Remember $\log(p/q) = \log p - \log q$ and V_f is constant)
8. Plan a graph using your V and t data which you would expect to be a straight line of gradient – k.
9. Plan a table of readings, and further columns, to provide values you will plot on your graph.
10. What must you do to ensure that the volume readings are all taken at the same pressure?
11. What could you do to ensure the rate will be unaffected by any change in temperature?
12. Which part of the procedure allows you to obtain a value for V_f without waiting for ever?

Post-laboratory work/questions

1. Complete your table. Plot your graph. Does it confirm first order kinetics?
2. Deduce the value and units of k. Compare your value with those of other students.
3. Outline how you could use this general method to find E_a for the reaction.
4. What, if anything, do your results tell you about the mechanism of the reaction?

Comments
The pre-laboratory exercise leads students through the many conceptual steps between the rate equation and the procedure. Extensions to study precision or effect of pH or counter ion are possible. Connections are possible to organic courses — eg by making an azo dye by reaction of some spare PhN_2^+ with a different nucleophile such as phenoxide.

Source
University of Glasgow

19 The nitration of methyl benzoate and monitoring by thin-layer chromatography

Level	*Two*	*Aims*	
Prior information	*Lectures on substitution reactions of benzenes and a manual with detailed instructions.*	*To carry out an aromatic substitution; to exemplify lecture material; to think about rates and site selection, procedure and physical properties.*	

Pre-laboratory work/questions

1. Draw out in full the formula for methyl benzoate. Which atoms lie in a plane?
2. What effect will the carbonyl oxygen have on the π electrons of the neighbouring aromatic ring?
3. Will methyl benzoate be a better or worse nucleophile than benzene? Give your reasons.
4. What implications does this have for the choice of electrophilic reagent?
5. What is the role of the sulfuric acid in the mechanism?
6. Write an equation for the reaction of nitric and sulfuric acids to give the ion NO_2^+.
7. What hazards are involved in handling the acids and how does the procedure reflect these? Sulfuric acid is also the solvent for the reaction. How will it be separated from the product?
8. Express the amounts of all the reagents in moles and work out the weight of methyl 3-nitrobenzoate that corresponds to a 100% yield.

Post-laboratory work/questions

1. Report the yield and mp of the crude and recrystallised product.
2. Write the mechanism for the substitution reaction indicating which of the two steps is rate-determining.
3. Draw your tlc plate and comment on the relative R_f values of reagent and product.
4. Was there any evidence from the tlc plate of dinitration? What theoretical reasoning accords with this? Is there any evidence of isomeric mononitroesters? What theoretical reasoning accords with this?
5. Why was it important to break up the lumps of precipitate in the water?
6. What risk would there be if ethanol were chosen as solvent for the recrystallisation rather than methanol?
7. Suggest why the product is solid and slightly yellow whereas the reactant was not. What modifications of this procedure might be needed for the mononitration of 1,4-dimethylbenzene?

Comment
Covers hazards, quantities, yield, procedure and work-up, concepts of delocalisation, catalysis and reactivity. Deductions from observation, questions on procedure, checks on understanding of mechanistic ideas, polarity, ultraviolet absorption. Some lateral thinking required.

Source
Heriot-Watt University and University of Glasgow.

20 The condensation of acetone with excess benzaldehyde in the presence of sodium hydroxide. Interpretation of spectra

Level	*Two*	**Aims**	
Prior information	*Lectures on aldol reaction and basic spectroscopy. Formulae and instructions.*	*To carry out and understand an aldol condensation; to use spectroscopic evidence for structure; to use tlc and mp as evidence of purity; and to think about procedure and mechanisms.*	

Pre-laboratory work/questions

1. Draw the delocalised carboxylate anion formed on deprotonation of an acid and the enolate anion similarly formed on deprotonation of a ketone. Look up in your textbook the pK_a values for acids and ketones and for water and methane and suggest reasons for the differences.

2. Use the Henderson equation to predict the results of adding a little of an acid to water at pH 14 and of adding a little ketone to water at pH 14.

3. Write out the stages of the reaction of the enolate ion of acetone with benzaldehyde (addition, proton transfers, expulsion of hydroxide ion) leading to 4-phenylbut-3-en-2-one.

4. Why does the benzaldehyde not give rise to an enolate ion?

5. Write out the balanced equation for condensation of acetone with two moles of benzaldehyde to give diphenylpentadienone (dibenzylideneacetone). What is the role of the sodium hydroxide?

6. Work out the number of moles of each reactant you are told to use and suggest a reason for the proportions. What implications are there for isolation of pure product?

7. How might infrared and ^{1}H NMR data help to establish the geometry of the double bonds of the product?

Post-laboratory work/questions

1. Calculate the theoretical and actual yields of dienone.

2. Suggest a different dienone that might have been formed by an alternative reaction of acetone and suggest why very little of it was formed.

3. What evidence had you for the purity of your product? Comment on the relative R_f values of the dienone and benzaldehyde.

4. Suggest why your product is yellow and why it is not very soluble in methanol.

5. The spectra will show bands roughly as follows. Find the exact values. Explain them and their significance. Infrared bands at about 1660 cm^{-1} and 980 cm^{-1} and an ^{1}H NMR doublet at about δ 8.

Comment
Connects with other courses. Analyses mechanism, procedure and physical properties.

Source
Heriot-Watt University, Lancaster University and University of Glasgow.

The place of pre-laboratory exercises

Why laboratory work in chemistry?

Laboratory work has been a part of undergraduate programmes in chemistry for nearly 200 years. Today, laboratory work is generally agreed to be an essential part of any chemistry course, although the purposes, methodology and time allocation have been the subject of much debate (S.W. Bennett, *Educ. Chem.*, 2000, **37**, 49).

The recent QAA benchmarking document (**http://www.qaa.ac.uk**) lists among the essential elements of a degree course in chemistry:

▼ *practical skills (including safety, hazards, risk assessment, procedures, instruments, observation and measurement, evaluation and interpretation of results, planning and selection of methods);*
▼ *transferable skills (including team working, organisation, time management, communication, presentation, information retrieval, data processing, numeracy, ICT, designing strategies, problem solving); and*
▼ *intellectual stimulation, connections with the 'real world', raising enthusiasm for chemistry.*

Most of these will be – and perhaps can only be – achieved in laboratories or in laboratory-related activities.

Others (M.Pickering, *J.Chem. Educ.*, 1989, **66**, 845) have highlighted the educational value of decision taking and discovery by students. It could be argued that laboratories might illustrate scientific method, might build confidence and might improve understanding. They of course allow students to see reactions, substances and effects, and can encourage student-student and student-staff interactions.

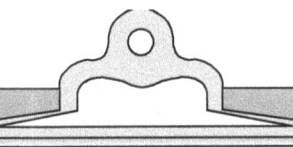

**Practical work in undergraduate courses
Some reasons for its inclusion.**

- Illustrating key concepts
- Seeing things for 'real'
- Introducing equipment
- Training in specific practical skills and safety
- Teaching experimental design
- Developing observational skills
- Developing deduction and interpretation skills
- Developing team working skills
- Showing how theory arises from experimentation
- Reporting, presenting, data analysis and discussion
- Developing time management skills
- Enhancing motivation and building confidence
- Developing problem solving skills

In surveys of the practical work being conducted in universities throughout the UK, M.A.M Meister and R. Maskill, *Int. J. Sci. Educ.*, 1995, **17**, 575 showed that, while there is considerable diversity in detail, there is also much common ground, with similar types of experimental procedures being introduced in many courses. Since the compilation of their reports, it seems that the situation has not changed dramatically although there are now some simpler experiments for less well qualified applicants and there is an increased diversity in how laboratories are run and what is expected of them. However, the demands and pressures on students and their teachers have increased considerably.

Pressures on laboratory work

The past two decades have seen a quite remarkable pressure on higher education in the UK, with increasing numbers and falling unit resource levels. For laboratory work in particular, the pressures of increasing numbers of students coupled with restrictions on manpower, materials, equipment and contact hours have been significant.

Staff often feel that students come to the laboratories ill prepared and seem to learn little from their laboratory time. On the other side of the fence, students may find few connections between laboratories, lectures and assessments and may find prescriptive closed procedures – where the outcome is already known – to be boring. In fact some school laboratory work is more open-ended and may even be based on an investigative approach. They may feel that the mark credit is not worth the time and that there is no other gain. Sometimes, the assessment encourages corner cutting or copying rather than thought or effort.

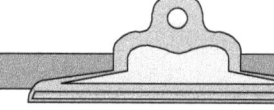

Practical work in undergraduate courses – modern pressures

- Cost of materials and equipment
- Safety issues and disposal of chemicals
- Staff and demonstrator costs
- Lack of student preparation (due, partly, to outside remunerative work)
- School experiences are very different (and entry levels are more variable)
- Assessment: what are we rewarding?
- Is the credit given worth the effort?

There is evidence that students are often overwhelmed by the information overload in the laboratory and resort to unthinking slavish performance – *eg* A.H. Johnstone, *J. Chem. Educ.*, 1997, **74**, 262. They may get tangled up with misconceptions, language, jargon, symbolism or representations.

Achievement by students regularly falls short of staff aspirations.

Coping strategies

With such pressures, inventive and motivated teachers have changed procedures and created support systems. For example, laboratory work has been changed to involve problem-based work, often with student collaboration, while laboratory manuals have been improved.

Some coping strategies

Using problem-based laboratories

Student cooperation in laboratories

Providing information on apparatus, instruments and procedures: by videos on CD, on WebCT or by using digital photos and text on CD

Computer programmes for simulations, practicing calculations, data handling and presentation, exploring of quantitative relationships, or testing knowledge

Preparation, support and practice in advance of the laboratory can take several forms and some of these allow repetition as often as wished in students' own time. Information on apparatus, instruments and procedures can be provided by videos, on CD (**http://www.cmf-v.com**), on WebCT (*eg* G. M. McKelvy, *Uni.Chem. Educ.*, 2000, **4**, 46) or by digital photographs and text on CD. Computer programmes for simulation, practicing calculations, data handling and presentation, exploring quantitative relationships, and tests of knowledge have been developed (*eg* J. Tomlinson, P. O'Brien and C. J. Garratt, *J. Sci. Educ.*, 2000, **1**, 100; D. Brattan, D. Mason and A. J. Rest, *Uni. Chem. Educ.*, 1999, **3**, 59; B. S. Nicholls, *Uni. Chem. Educ.*, 1999, **3**, 22.).

Many laboratory classes begin with a talk by a demonstrator in the laboratory that can establish rapport, identify locations and hazards, give advice, and refresh existing knowledge. However, this consumes laboratory time and is given once. It may vary in quality from day to day and, inevitably, it depends on the demonstrator's communication skills and grasp of the aims and priorities.

All these devices are valuable in preparing the minds of the students, reducing the sudden overload of information, laying the groundwork for the laboratory experience and report, and improving learning.

Facilitating learning

'If I had to reduce all of educational psychology to just one principle, I would say this: the most important single factor influencing learning is what the learner already knows. Ascertain this and teach him accordingly'. (Ausubel, 1968)

This statement, based on an accumulation of evidence, has been shown again and again to be extremely important in all learning and can be illustrated for chemistry (G. Sirhan *et al.*, *Uni. Chem. Educ*, 1999, **3**, 43-45). Much is now known about how sound learning is achieved.

▼ Students build their new learning on what they already know and understand.

▼ Learners can only handle a limited amount at any one time (working memory space limitations).

▼ There is a limit to the rate of processing of incoming information (to allow connections and storage).

▼ Feedback, discussion, application and reassurance are needed.

▼ Learners need occasionally to have the chance to make suggestions, to propose theories or strategies, to explore, to create and to present their distilled knowledge and understanding.
(*eg* D.P. Ausubel *et al.*, in *Educational psychology: a cognitive view*. London: Holt, Reinhart and Winston, 1978; A.H. Johnstone, *Cerapie*, 2000, **1**, 9–15). (http://www.uoi.gr/conf_sem/cerapie/2000_January/056johnstonef.html)

Pre-laboratory exercises

To gain the maximum benefit from time in the laboratory in chemistry, preparation by the learner is vitally important. For laboratory work to be successful, there must be a clear recognition by both students and instructors of the existing knowledge and experience base which the student possesses. In the light of this, instructors must clearly define for themselves the educational aims of the experiment: what the students should gain from it. Preferably, at least some of these aims should be made clear to the students. Even a short time in preparation can enormously improve understanding of procedures and equipment and interpretation of results.

This report discusses paper based pre-laboratory exercises, while recognising that there are other ways to prepare students for practical work. These exercises can be short but challenging. They prepare the learners' minds for what they will face in the laboratory, enabling students to gain much more from the time spent there. Too often, in laboratories, students are faced with an unfamiliar or forgotten theoretical background to the experiment. They may also meet unfamiliar equipment or instruments, wordy instructions that frequently have to be followed like a recipe, coupled to verbal instructions from demonstrators that can mix confusion with sound advice.

An early casualty is thought: students are so concerned about carrying out the experiment and gathering the necessary data that all thought of what is being done, why it is being done, and the significance of the experiment is left until the report writing stage. Pressure to hand in a polished report with a 'correct' answer further hinders deeper thought while it is impossible to go back to try the experiment again to modify procedures or to think about why the experiment is being done in the first place. In such circumstances, students adopt all kinds of strategies to cope with the mental overload – *eg* they might focus on certain tasks ignoring others.

Pre-laboratory exercises can allow the necessary background knowledge to be revisited, for experimental techniques to be introduced, for the significance of experiments and their design to be explored and for the introduction of important questions that can be addressed during or after the laboratory time. Post-laboratory questions and reports can be linked to these pre-laboratory exercises to make the laboratory experience more of a complete whole.

Pre-laboratory activities can assist where:

the knowledge base is inadequate;

lectures and laboratories are not integrated;

information overload overwhelms students;

laboratory manual and demonstrator might diverge;

undergraduate laboratories are manual led (less popular)while school practical work was more open-ended (popular); and

students would otherwise have no constructive role or ownership.

Laboratory effectiveness and pre-laboratory exercises

In various studies of the effectiveness of laboratory work – *eg* A.H. Johnstone and K.L. Letton, *Educ. Chem.*, 1990, **27**, 9–11 – it is clear that many of the aims set by the designers are not being fulfilled very well. Given the resource demand of practical work, this is a matter of concern. However, in studies where pre-laboratory activities have been employed (A.H. Johnstone, R.T. Sleet and J.F. Vianna, *Studs. Higher Educ.*, 1994, **19**, 77; A.H. Johnstone, A. Watt and T.U. Zaman, *Phys. Educ.*, 1998, **33**, 22), there is clear evidence that learning has increased and motivation has been enhanced. Together with post-laboratory activities, laboratory experiences for undergraduates can be enriched.

Designers of university laboratory courses might, on reflection, agree that their students experience a continuum of learning experiences. These might include: previous school work, chemistry lectures, courses in parallel subjects, reading, the internet, workshops, pre-laboratory activities, laboratory experiments, recording, interpreting, reporting, extending, interactive questioning, assessment and feedback, revision, and examinations. In considering pre-laboratory exercises, as well as any other material, it is important to design material and methods to take account of this continuum and the requirements for effective learning.

The aims of laboratory work

Articles, papers and laboratory manuals reveal the wide diversity of aims that have been listed for undergraduate chemistry laboratory work. Perhaps, these can be grouped into six broad overlapping themes.

Practical skills

Students may gain the kinds of skills that are widely used by chemists, including safety as well as the handling of equipment and instruments.

Preparing for the world of work

Only about 30% of chemistry undergraduates proceed on to 'chemical' jobs but there is the broader preparation for the world of work including personal skills as well as practical skills such as experimental design, managing time and risk, and the nature of precision. The world of work will be a place where problems have to be faced, analysed and solved and laboratory work provides considerable scope for experience in practical problem solving.

The methods of science

Students have opportunities to see something of the way science operates as it seeks to gain answers from the physical world by means of the interpretation of experimental data. There are opportunities to discover, to explore, to confirm, to observe, to report, to interpret and to challenge.

Aims of practical work

Making chemistry real

Laboratory work can make chemistry 'come alive', allowing students to see, touch and handle chemicals and equipment, to see how data is gathered, to see how theoretical models can be tested.

Personal skills

There are often opportunities for team working, planning, time management, discussion and debate. Good laboratory experiences can have a positive effect on student attitudes and motivation. Success leads to confidence and this frequently leads to positive attitudes towards chemistry, with a stronger motivation to move on to more demanding tasks.

Intellectual skills

Laboratory work can offer great opportunities for posing questions and understanding theoretical models when applied to the real world. Ideas, facts and rationale can be brought together, the skills of data analysis and interpretation can be developed. Observations can be made, explored and extended. The models of chemistry can be made more tangible.

Pre-laboratory exercises in use today

During the year 2000, a brief questionnaire was sent to all (83) chemistry departments in the UK and Ireland asking what forms of preparation for laboratories at level 1 were in use. The 47 replies represent about 60% of departments offering chemistry degrees but the responses may apply only to some courses and the survey is inevitably partial. The replies suggest that

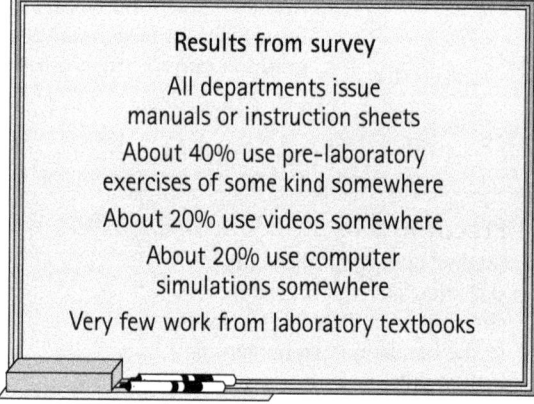

Results from survey

All departments issue manuals or instruction sheets

About 40% use pre-laboratory exercises of some kind somewhere

About 20% use videos somewhere

About 20% use computer simulations somewhere

Very few work from laboratory textbooks

while all departments issue manuals or instruction sheets, about 40% use some kind of pre-laboratory exercise to some extent. Responses to the questionnaire also indicated huge variations in class sizes, in the sophistication of experiments at nominally the same level, and in requirements in the way of post-laboratory reports or questions. Clearly there is a wide variety of aims and strategies among UK universities. They also confirmed that staff assessment of student achievement in various skills fell much below their assessment of the importance of these skills.

While quite a number of UK departments use some form of pre-laboratory exercises, correspondence with colleagues overseas suggests that there is little use in Australia or the continent of Europe and there are only a few reports of use in the literature from Canada and the US. However, some laboratory textbooks now include them (*eg* J.M.Bauer and M.M. Bloomfield, *Laboratory experiments for chemistry and the living organism.* Chichester: Wiley, 1992. H.R. Hunt, T.F.Block and G. M. McKelvy, *Laboratory experiments for general chemistry.* Fort Worth: Saunders, 1998).

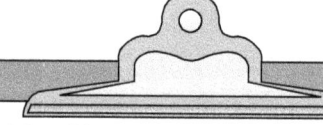

A bewildering variety of requirements
Most manuals define the procedure and quantities completely.
Write-up may require:
- A formal account with conclusion or
- brief results with questions to check understanding

Students are sometimes asked to think about:
- Procedures
- Precision and reproducibility
- Interpretation, comparison, alternatives, extensions, significance

Sometimes they are not asked to think at all.

Marking may be:
- By people or computer
- Immediate and face to face or
- Delayed and anonymous

The authors and users of pre-laboratory exercises and activities clearly believe they are useful and helpful. Many have had supportive comments from students and there is some evidence of effectiveness in changing attitudes and from performance and examination results. Following the questionnaire, samples of pre-laboratory exercises with manuals were requested from all the known users. (All those who sent materials and allowed us to reproduce or modify parts of them are gratefully acknowledged.)

As might be expected, the collection revealed great variations in the organisation and content of the manuals and in what was wanted by way of post-laboratory reports or interpretation of results. Some manuals make clear the aims of the work and the skills to be developed. Surprisingly, some don't!

Many manuals provide all the background theory [perhaps not yet given in lectures], while some involve video or appendices to show details of equipment or instrument techniques and keep operating instructions to a brief check-list while requiring (by means of a pre-laboratory exercise) that students understand what is happening and why. The choice of manual style and methods will be affected by the aims of the authors, the level of the class, the manpower and hardware resources and other factors. There is no 'best way' – only a 'best way in this case'.

The collection revealed the great range of styles, content and length among existing pre-laboratory exercises and the ingenuity, awareness and lateral thinking of some authors. Most of the pre-laboratory exercises received were at level 1 but there were a few for level 2 or 3. It is understandable that preparation of students' minds is seen to be most needed in their first year after school. However, for Honours experimental project work the requirement to find out about background and think about strategy, methodology or interpretation before an experiment is started is universal and seen as essential. It can be argued that, in suitably modified form, such preparation is appropriate at **all** levels.

It is clear that, at least in some courses, the controlled predictable 'experiments' that were almost universal 10 years ago are being adapted or replaced. This change may be driven or may be inspired by the changing range of students, the changing technology, the different aims and requirements and a growing awareness that **it is** possible to improve understanding, involvement and motivation.

What might be included in pre-laboratory exercises?

Assuming that the manual states the aims of the experiment and, perhaps, the skills to be practised along with some background and instruction on procedure and technique, what can a pre-laboratory exercise do to prepare students' minds? From the manuals that were collected and from discussions with colleagues, a rough alphabetic list was drawn up. This shows the range of topics and themes that might be important in pre-laboratory exercises.

▼ *Apparatus, glassware, instruments, handling*
▼ *Calculations, concentration, unit conversions*
▼ *Equations, reactions, selectivity, catalysis, physical principles, concepts*
▼ *Explaining, thinking out, applying theory, understanding of theory or procedure*
▼ *Facts, formulae, data, physical constants, literature information, searching and abstracting, definitions*
▼ *Group work, compilation or comparison of results, discussion*
▼ *ICT skills related to searching, presentation, data handling*
▼ *Jargon, nomenclature, terminology*
▼ *Moles, excess reagents, yield calculations*
▼ *Planning of procedure (flowchart), quantities, timing, temperature, completion, work-up, purification*
▼ *Planning of recording, tables, graphs, using real data, report, deductions, interpretation, diagrams*
▼ *Precision, calibration, reproducibility*
▼ *Predictions, correlations, synthesis of ideas, discovery*
▼ *Reassurance, motivation, ownership, real world relevance, sources of materials, costs*
▼ *Safety and disposal*

Using this list, it is possible to draw up a series of 12 themes and topics that might be included in a pre-laboratory exercise.

Themes and topics for pre-laboratory exercises

1. *Establish the context and stepwise connections between lectures and theory, procedure and reagents*

2. *Demand recall of existing knowledge*

3. *Check understanding and use of terms, symbols, units, mechanisms*

4. *Check the instructions/appendix/textbook/video has been read and understood*

5. *Make clear the significance of the experiment or its connections with the real world*

6. *Practise operations like moles or yield calculations, graphs, drawings, computer use*

7. *Require literature searching, references, safety data, sources, disposal*

8. *Help understanding of the given procedure and data handling*

9. *Require planning of the procedure and data handling*

10. *Invite thought about why we are doing this, why in this way, or what we might make of the outcome*

11. *Establish confidence and competence, improve motivation and ownership*

12. *Organise package and connect knowledge.*

It is important to see the laboratory experience as a total experience, involving pre-laboratory exercises, the actual laboratory as well as post-laboratory activities. Possible activities which might follow the laboratory are now considered.

Post-laboratory exercises

After the experiment, the students will report and, it is hoped, reflect on what they have been doing. This will include activities like data handling and writing conclusions but many other useful activities can be included.

▼ *Interpretation of results and observations during the experiment (both intentional and incidental)*

▼ *Comparison among the class, with literature, with predictions, with other methods or other reagents*

▼ *Exploration of implications, applications and extensions*

▼ *Re-examination of the procedure, alternatives or improvements*

▼ *Discussion with other students or with demonstrators*

The idea needs to be instilled steadily that 'lab-work is a thinking task supported by laboratory equipment'. A gradual paced process of hearing, reading, thinking, doing, seeing and thinking again offers the best recipe for effective learning in and from the laboratory.

'*...lab-work is a thinking task supported by laboratory equipment.*'
A. Berry et al, *Australian Teachers' Journal*, 1999, 45(1), 27.

Writing pre-laboratory and post-laboratory exercises

Pre-laboratory written exercises that students present before they start an experiment offer one way of:

▼ *preparing students' minds;*

▼ *checking their reading and understanding;*

▼ *helping them to understand the logic, procedure and significance;*

▼ *practising key skills;*

▼ *developing increased interest and motivation;*

▼ *extending the time they have to digest ideas; and*

▼ *making the time in the laboratory more productive.*

They need to be seen as part of a range of activities and of learning and testing situations. The same general principles that apply in developing any educational material apply when thinking of writing pre-laboratory and post-laboratory exercises. Some of these considerations are now discussed briefly.

Aims

It seems obvious that, before we design an experiment or exercise we need to be clear why we are doing it, and what knowledge, experience, understanding, practice, or skills it provides.

It probably helps if these aims and objectives are made clear to students and to demonstrators by the writers. It is all too easy for these three groups to have different conceptions of the purpose of the experiment.

It probably helps to make clear how these objectives will be reflected in the assessment or in later testing.

In the examples here, aims were given very briefly. More detail is probably desirable.

Level

The content, aims, format, length and wording of the pre-laboratory exercise depends on the level of the class and may vary from one university to another.

In the examples here, there is a rough gradation in terminology, length and complexity.

Prior knowledge

The writer will take into account the previous knowledge of the students – what they will know from school or earlier years, from other courses and, in particular from lectures, workshops, web searches and videos which were part of the course. There are problems, of course, if not all students have the same experience because of heterogeneous entry or variations of timing.

It helps to recognise and address likely sources of confusion - *eg* terminology, symbols and mathematical manipulations. Knowledge, from experience, of things that have caused difficulty when the experiment was run before provides excellent clues for what might usefully be included in the pre-laboratory exercise.

Example 18 was written this way. The experiment has been running for years with no pre-laboratory exercise. All the background facts and equations were provided in the manual even though much of this was in the previous year's lecture course. All the instructions about what to measure and what to do with the numbers are given. Students came to the laboratory having read the manual but understanding very little. Certain motivated demonstrators talked through the reasons and relationships between procedure, data, mathematics, graph, rate law and meaning. Others did not. The (unstated) aim seemed to be to assume a rate law and then prove you were right. Little thought was given to the strategy. Why can we not measure reagent concentration? How does volume of product relate to concentration of reagent? Why is a volume ratio numerically equal to a concentration ratio? What experimental controls ensure the validity of this statement?

Understanding the general gas law, the stoichiometry, the mathematical equivalences and mathematical manipulations, and the practical niceties are important outcomes but were not so identified. No use was made of the eventual numerical value of the rate constant. No connections were made with organic courses on diazonium ions or with interpreting the rate law. Maybe it was assumed students would do all that for themselves.

The real value of the experiment lies in understanding the strategy, the apparatus, the related mathematics and physical chemistry. The pre-laboratory questions suggested here were designed to help with that, and the post-laboratory invites thought about how the experiment could be extended and what conclusions can be drawn.

Assessment

Assessment credit may be necessary to encourage completion of pre-laboratory exercises or may be desirable in itself as part of overall assessment. It demands marking by demonstrators, preferably face to face, at the beginning of the laboratory, or adjustment of the exercise for computer marking. This takes time and manpower but there will be gains in terms of understanding and through adjustment of the discussion to the needs of individual students. This marking time might previously have been devoted to assessing the reports and writing feedback. The amount of credit might depend on the length, time required and coverage of the pre-laboratory exercise.

Pre- and post- questions

Usually reports or post-laboratory questions are longer than pre-laboratory exercises and can exploit, interpret and extend the experimental observations. Often questions about procedure or requirements such as drawings, mechanisms, or literature information might be in either. Sometimes the pre-laboratory exercise lays the groundwork for the report by planning the recording and manipulation of data or practising calculations or making estimates or hypotheses to be checked. It may draw attention to important observations, critical stages, or features of spectra or graphs to be reported afterwards.

Epilogue

It is important to see the pre-laboratory exercises, the laboratory performance and the post-laboratory report forming an integrated learning sequence. Together, they can fulfil some of the specific aims that are seen to be desirable for undergraduate chemistry courses and may assist in the development of chemistry graduates well equipped to move on to the work place.

The examples in this booklet illustrate some of the ingenuity and perception shown by writers of materials from various universities. There may well be more excellent material elsewhere. Hopefully the examples and discussion here will prove helpful to colleagues anxious to improve their students' experience, by giving the students more responsibility for their learning, and making the precious laboratory time as effective and fruitful as possible.

> *To change the experience, you don't need to change the experiment, just what you do with it.*

RS•C

This page has been intentionally left blank.